ACCELERATING CLIMATE AND DISASTER RESILIENCE AND LOW-CARBON DEVELOPMENT THROUGH THE COVID-19 RECOVERY

TECHNICAL NOTE

OCTOBER 2020

ADB

© 2018 Asian Development Bank
6 ADB Avenue, Mandaluyong City, 1550 Metro Manila, Philippines
Tel +63 2 632 4444; Fax +63 2 636 2444
www.adb.org

Some rights reserved. Published in 2020.

ISBN 978-92-9262-433-0 (print); 978-92-9262-434-7 (electronic); 978-92-9262-435-4 (ebook)
Publication Stock No. TCS200295-2
DOI: http://dx.doi.org/10.22617/TCS200295-2

The views expressed in this publication are those of the authors and do not necessarily reflect the views and policies of the Asian Development Bank (ADB) or its Board of Governors or the governments they represent.

ADB does not guarantee the accuracy of the data included in this publication and accepts no responsibility for any consequence of their use. The mention of specific companies or products of manufacturers does not imply that they are endorsed or recommended by ADB in preference to others of a similar nature that are not mentioned.

By making any designation of or reference to a particular territory or geographic area, or by using the term "country" in this document, ADB does not intend to make any judgments as to the legal or other status of any territory or area.

Please contact pubsmarketing@adb.org if you have questions or comments with respect to content, or if you wish to obtain copyright permission for your intended use that does not fall within these terms, or for permission to use the ADB logo.

Corrigenda to ADB publications may be found at http://www.adb.org/publications/corrigenda
This publication may be accessed online at http://dx.doi.org/10.22617/TIM189600-2

Notes:
In this publication, "$" refers to United States dollars unless otherwise stated.
ADB recognizes "China" as the People's Republic of China and "Vietnam" as Viet Nam.

Contents

Table and Figures   iv

Acknowledgments   v

Abbreviations   vi

Executive Summary   vii

I.   Introduction   1

II.   Issues   3

    A.   Recovery Planning Is an Opportunity to Promote Long-Term Transformation   3

    B.   Drivers for Integrating Climate and Resilience Considerations into the Recovery   3

    C.   Resilient and Low-Carbon Interventions in Support of Economic Recovery   8

III.   Opportunities for a Low-Carbon and Resilient Recovery   12

    A.   An Assessment Framework for Climate and Resilience Benefits   12

    B.   Opportunities for ADB to Support Its DMCs   15

Appendix: Climate- and Disaster-Resilient and Low-Carbon Recovery Interventions   20

References   24

Table and Figures

Tables

1   Framework for Assessing Low-Carbon and Resilient COVID-19 Recovery Interventions   14

Figures

1   Low-Carbon and Resilient Recovery Interventions Align with Strategy 2030   15
2   Strategy 2030 Operational Priority 3   17
3   ADB and Paris Alignment   19

Acknowledgments

This technical note was prepared by Kate Hughes, senior climate change specialist, Sustainable Development and Climate Change Department (SDCC), Asian Development Bank (ADB) and Arghya Sinha Roy, senior climate change specialist (adaptation), SDCC, ADB under the overall guidance of Preety Bhandari, director, Climate Change and Disaster Risk Management Division, SDCC, ADB, and with inputs from Charlotte Benson, principal disaster risk management specialist, SDCC, ADB. Ken Concepcion, senior operations assistant, SDCC, ADB, provided valuable coordination support. This technical note builds on the previous publication COVID-19 Recovery: A Pathway to a Low-Carbon and Resilient Future https://www.adb.org/sites/default/files/publication/625476/covid-19-recovery-low-carbon-resilient-future.pdf.

Paul Watkiss (consultant) provided inputs to the early draft of document. The document also benefited from discussions with, and comments received from Chen Chen, country director, Sri Lanka Resident Mission, ADB; Virender Kumar Duggal, principal climate change specialist, SDCC, ADB; Yukiko Ito, senior social development specialist, SDCC, ADB; Xuedu Lu, lead climate change specialist, and colleagues from East Asia Department, ADB; Mohd Sani Mohd Ismail, senior financial sector specialist, Southeast Asia Department, ADB; Nathan Rive, senior climate change specialist, Central and West Asia Department, ADB; Hanif Rahemtulla, principal public management specialist, SDCC, ADB; Eric Sidgwick, former country director, Viet Nam Resident Mission, ADB; Paolo Spantigati, country director, and colleagues from Armenia Resident Mission, ADB; Sonomi Tanaka, chief of gender equity thematic group, SDCC, ADB; Liping Zheng, advisor, and colleagues from South Asia Department, ADB; and consultants David Bloomgarden, Roberta Gerpacio, Rosa Perez, and Zonnibel Woods. The manuscript was edited by Mary Ann Asico. The infographics and layout were designed by Rocilyn Laccay.

Abbreviations

ADB	Asian Development Bank
CCS	carbon capture and storage
COVID-19	coronavirus disease
DMC	developing member country
DRR	disaster risk reduction
GDP	gross domestic product
GHG	greenhouse gas
IEA	International Energy Agency
LTS	long term strategies
NAP	national adaptation plan
NDC	nationally determined contribution
R&D	research and development
SDG	Sustainable Development Goal
TA	technical assistance

Executive Summary

The coronavirus disease (COVID-19) pandemic is a major global crisis requiring national, regional, and global intervention. Economic losses resulting from the crisis in Asia and the Pacific could reach $1.7 trillion–$2.5 trillion under different containment scenarios, according to the Asian Development Bank (ADB), and between 90 million and 400 million people in Asia and the Pacific could be pushed back into poverty, living on less than $3.20 a day. Governments and international financial institutions are expected to respond with unprecedented funding of at least $10 trillion. The developing member countries (DMCs) of ADB have been under intense pressure to implement emergency response measures that address the immediate public health risk, and ease its wider effects through social protection and business support, while also limiting the scale of economic damage. ADB is deploying a $20 billion response package to help the DMCs deal with the immediate impact of COVID-19.

To move forward, the DMCs must plan their recovery over the medium and long-term. In the medium term, countries are expected to focus on economic stimulus measures to restart their economies and reengage their workforce, as well as on accompanying reforms, such as changes in monetary policy. Over the long term, considerations must center on the impact of COVID-19 on the countries' development pathway, their vision for the future, and the steps they must take, including transforming systems, institutions, and policies, to achieve that vision. The ability to develop and implement a COVID-19 recovery plan will vary between countries, and should be viewed in the context of what could be an even deeper global recession. Low- and middle-income countries will have less fiscal space than high-income countries to go beyond their immediate emergency response. Given the magnitude and uncertain duration of COVID-19, a large number of countries will need substantial international support in addition to emergency COVID-19 support they have already received.

A strong call has been made to "build back better" and for the COVID-19 recovery to be "green," taking advantage of the massive expected global stimulus to tackle the ongoing climate crisis, invest in a more sustainable and resilient future, and deliver on key international agreements including the Paris Agreement, the Sendai Framework for Disaster Risk Reduction and the Sustainable Development Goals (SDGs). Having neither the time nor the financial resources to deal separately with the COVID-19 and climate crises, countries must avoid letting concerns about the economic impact of COVID-19 override the urgency of taking climate action. The lives and livelihoods of millions of people, especially the poor and vulnerable, are already affected by climate change, and progress in sustainable development is being undermined. Analysis done in 2019 estimated that climate change could push an additional 100 million people globally into poverty by 2030. By 2050, it could cut agricultural yield by up to 30%, and add more than $1 trillion each year to adaptation costs in urban areas.

The restraints imposed by governments on mobility, as well as the drop in economic activity caused by the pandemic, have reduced global greenhouse-gas (GHG) emissions in the short term (estimates vary between 4% and 11% reduction globally in 2020). To reach the goals of the Paris Agreement, this level of reduction would need to be sustained year on year until 2030. However, GHG emissions are likely to rebound fairly quickly, as conditions after the global financial crisis showed. The reduction in GHG emissions in 2020 has also been obtained through unsustainable methods—severely restricting movement and economic activity—as well as with unacceptable human impact including illness and loss of life. Sustaining the reduction will depend greatly on how countries choose to pursue their recovery and requires transformative changes to move away from fossil fuel dependent economies.

In parallel, COVID-19 has exposed systemic risks across sectors and themes—health, climate change, disasters—and made very clear the need to deal with the underlying drivers of vulnerability, including poverty, limited coverage of social safety net programs, weak health systems, social exclusion and structural gender inequality, among others. The need for a fundamental re-alignment of basic systems with the values underpinning the SDGs is clear. The lessons learned and the insights gained so far from the COVID-19 experience have made a strong case for risk-informed decision approaches, and have led to broader recognition that risks across different sectors and themes must be addressed holistically at a systems level.

The scale of the crisis and the associated size of the response means that decisions governments make now will influence systems, create assets, and define development directions well into the future. Countries can use this opportunity, and the expected stimulus, to reorient their economies toward a more strategic, low-carbon trajectory while at the same time addressing underlying vulnerabilities, and improving the climate and disaster resilience of communities and sectors. Recovery plans need to go beyond the immediate emergency response and strive to lift economies out of recession in the medium term, while laying a solid foundation for long-term prosperity. However, under great uncertainty and pressure, countries may find it challenging to consider long-term goals over short-term payoffs. To support decision making, they must understand clearly the long-term economic and social benefits they can derive from pursuing a low-carbon and climate-resilient recovery, as well as how the recovery interacts with, and can support realization of, nationally determined contributions and long-term strategies under the Paris Agreement. Countries need to adopt a transformative mindset to truly prepare themselves for a low-carbon and resilient future, and this needs to apply to how they navigate their recovery from the pandemic. This includes making the right type of investments, but also supporting policies to provide market signals and economic incentives to sustain behavioral changes beyond the stimulus period.

A low-carbon, resilient pathway will not necessarily raise the level of investment spending on the recovery effort. Rather, it will require a shift in the nature of interventions. Analysis done by the Global Commission on the Economy and Climate in 2014 established that strong climate action by countries between 2018 and 2030 could, by 2030, generate over 65 million new low-carbon jobs, deliver at least $26 trillion in net global economic benefits, and avoid 700,000 premature deaths from air pollution. There is a wide range of interventions that can deliver strong economic and social benefits to achieve recovery goals, and at the same time address climate change and improve resilience. Ample analysis exists to support this, as outlined in this technical note. For example

- building insulation retrofits or clean energy infrastructure are labor intensive in the early stages, can deliver high multipliers and have high returns over the long-term by driving down the cost of the clean energy transition. One widely cited model suggests that every $1 million in spending generates 7.49 full-time jobs in renewables infrastructure and 7.72 full-time jobs in energy efficiency but only 2.65 full-time jobs in fossil fuels;

- recent economic analysis by the Climate Council of Australia (as part of its proposed Clean Job Plan), estimates that investment in pilot-scale hydrogen facilities would unlock A$4 for every dollar of public investment; utility-scale renewable energy would elicit A$3 of every dollar invested; and electric vehicle infrastructure, improving the collection and processing of organic waste, and community scale energy and storage would return A$2 for every dollar invested; and

- recent estimates suggest that the net benefit of investing in more resilient infrastructure in low- and middle-income countries is $4 in benefit for each $1 invested.

There are widely varied options for spending stimulus funds to create jobs and jump-start economic growth. Not all of these options will support a long-term vision of a low-carbon and resilient future, but the list of possible recovery interventions that do is long (see the Appendix). It includes few "new" ideas. Some interventions are already being carried out in a number of countries; others, after being assessed in the past, may have been turned down or deferred. However, new drivers in the current pandemic situation, including changed perceptions of risk, could boost the adoption of these measures, and the mobilization of stimulus funds could present opportunities to speed up implementation.

ADB has developed this technical note to help the DMCs accelerate climate and disaster resilience and low-carbon development through the design of COVID-19 interventions in their medium-term recovery and long-term transformation phases. The technical note describes how a large number of the recovery interventions that are desirable with respect to delivering stimulus benefits can also improve climate and disaster resilience or support low-carbon development. When the package includes measures directed at inequalities and vulnerabilities exposed by the pandemic, countries gain the opportunity to drive a sustainable economic recovery, reduce inequality, build resilience and deliver the SDGs.

Different countries will have different criteria for recovery interventions, depending on many factors such as a country's main economic sectors, COVID-19 experience, recovery budget, and access to international support. However, as suggested in the literature analyzed in this technical note, interventions that will enable a "good" economic recovery from COVID-19 share common characteristics with economic stimulus initiatives that were successful in the past. These interventions may be designed

- to generate jobs and stimulate economic activity;

- with a short implementation timeline, so that the stimulus funds can be quickly deployed and people can get back to work in a timely manner (and possibly with less need for safety nets);

- to be labor-intensive, particularly in the early stages;

- to promote skills development, including re-skilling unemployed workers from sectors that have been heavily affected by COVID-19, or preparing for a low-carbon future;

- to have a strong supply chain, which could include considerations such as more localized or more diversified sourcing; and

- to have high economic multipliers.

For a low-carbon, resilient recovery, these criteria still apply, but they should be supplemented with other essential features that support a "build back better" approach, consider sustainability of benefits and the impact on the medium- and long-term development pathway.

ADB has developed a framework for assessing low-carbon and resilient recovery interventions (see table in Section III, page 17) to assist its DMCs in evaluating low-carbon and climate- and disaster-resilient interventions for recovery. The framework provides a structured process for evaluating, and comparing, the potential of the interventions to achieve the intended recovery. Decision-makers can use the framework, tailored to reflect each DMC's specific circumstances and recovery objectives, to select and prioritize a package of interventions that will collectively achieve recovery, climate, and resilience objectives. A rapid initial assessment of potential interventions based on a qualitative assessment of measures can be made (as shown in the table), or the framework can be developed further to allow deeper analysis, including, for example, a separate assessment for each phase of recovery or a quantitative assessment of the performance of recovery measures.

Governments in many countries have limited capacity to deal with COVID-19. Their revenues having been curtailed by the crisis, they have less fiscal space for the necessary reforms and investment. ADB's support can be critical to recovery-related planning and implementation by its DMCs. ADB can support the DMCs to use the assessment framework presented in this technical note, to support the design of recovery plans, including alignment with the objectives and processes under the Paris Agreement and other key international agreements. ADB can also support better access to opportunities under current ADB projects and programs, support in planning institutional and policy reforms, and integration of these considerations into country partnership strategies, as well as mobilization of finance to fund the recovery.

I. Introduction

1.   **The coronavirus disease (COVID-19) pandemic is a major global health, social, and economic crisis.** The developing member countries (DMCs) of the Asian Development Bank (ADB) have been under intense pressure to implement emergency response measures that address the immediate public health risk, and mitigate the wider impact of the crisis, through social protection and business support. To move forward, the DMCs must plan their recovery over the medium- and long-term, while taking into account the ongoing uncertainties related to COVID-19 and its impact. According to ADB estimates (ADB 2020b), economic losses in Asia and the Pacific could reach $1.7 trillion–$2.5 trillion under different containment scenarios, and between 90 million and 400 million people in Asia and the Pacific could be pushed back into poverty, living on less than $3.20 a day. Governments and international finance institutions are set to mobilize unprecedented funding—at least $10 trillion—in response (GCEC 2020). ADB is supporting its DMCs in responding to the immediate impact of COVID-19 through a $20 billion response package.[1] As the DMCs move forward, they will need additional assistance, including support for the private sector (ADB 2020a).

2.   **The COVID-19 recovery provides an opportunity to promote resilient and low-carbon development.** A strong call has been made to "build back better" and for the COVID-19 recovery to be "green,"[2] taking advantage of the massive expected stimulus to make investments, and accompanying policy changes, that build resilience, tackle the ongoing climate crisis, address inequality, and deliver on key international agreements including the Paris Agreement, the Sendai Framework for Disaster Risk Reduction and the Sustainable Development Goals (SDGs). COVID-19 has exposed systemic risks across sectors and themes—health, climate change, disasters—and highlighted the need to address the underlying drivers of vulnerability, including poverty, inequality, limited social safety nets, weak health systems, and social exclusion and structural gender inequality, among others. The scale of the crisis and the accompanying response means that decisions governments make now will influence systems, create assets, and define development directions well into the future. The DMCs can therefore use the recovery as an opportunity to reorient their economies toward a more strategic low-carbon trajectory, while simultaneously addressing underlying vulnerabilities, improving the climate and disaster resilience of communities and sectors, and accelerating delivery of the SDGs. There are a large number of interventions that can achieve recovery goals by delivering strong economic and social benefits, and at the same time provide climate and resilience benefits. But countries need to adopt a transformative mindset to prepare themselves for a low-carbon and resilient future, and apply this to how they navigate their recovery from the pandemic.

3.   **In responding to the crisis, governments have so far taken limited advantage of this opportunity for more resilient and low-carbon development**, particularly if their economies depend on environmentally intensive sectors such as coal mining, fossil fuel production, commercial agriculture, and logging. One study examined 17 stimulus packages totaling around $3.5 trillion and found that governments have largely ignored

---

[1]   ADB has approved a number of technical assistance (TA) and Asia Pacific Disaster Response Fund grants, as well as various nonsovereign operations, and worked with its DMCs to develop solutions, including emergency assistance loans and grants, to mitigate the immediate impact of the crisis. On 18 March 2020, ADB announced a $6.5 billion package to address the immediate needs of its DMCs as they respond to the COVID-19 pandemic. On 13 April 2020, ADB expanded its response to a total of $20 billion by making available up to $13 billion in additional regular ordinary capital resources to finance countercyclical expenditures, and providing additional grant and TA resources.

[2]   A variety of terms have been used in this context, e.g., "green recovery," "green reset," "green new deal," and "building back better." The term is also used to cover a broad range of issues and measures. This technical note is specifically focused on climate and disaster resilience, as well as low-carbon development, which can be considered a subset of a green recovery.

the broader sustainability and resilience impact of their actions (Vivid Economics 2020). There are exceptions. For example, the recovery measures announced so far by the People's Republic of China include some "green" elements, such as investments in high-speed rail, electric vehicle charging infrastructure, and ultra-high-voltage power transmission (Gosens and Jotzo 2020). "Green" remodeling of public facilities, buildings, housing, and manufacturing bases is part of the Republic of Korea's recovery plan (Regalado 2020). India has allocated $780 million toward an afforestation program designed to stimulate the rural and semi-urban economy while providing essential ecosystem benefits (Vivid Economics 2020). These measures, however, tend to be stand-alone elements of recovery plans, rather than being part of a unified "green" recovery approach. Furthermore, the benefits are often overshadowed as the packages also contain so-called "brown"[3] measures (usually in larger proportion than green measures). Mostly, governments have tended to favor business-as-usual approaches, providing large rescue packages for carbon-intensive industries, or worse, rolling back existing environmental standards. In doing so, they could lock in decades of high-carbon and unsustainable development, deepening existing inequalities (Levy, Brandon, and Studart 2020). In the Asia and Pacific region, recent analysis found that several countries have taken decisions that lock them into a trajectory of even higher GHG emissions than what they were on before COVID-19 occurred (Carnell, Sakpal, Pang, Mapa, and Patterson 2020).

4.   **This technical note provides guidance on accelerating climate and disaster resilience and low-carbon development through the COVID-19 recovery.** ADB has developed this technical note to help its DMCs accelerate climate and disaster resilience and low-carbon development through the design of their medium- to long-term COVID-19 recovery plans. The technical note describes how a large number of the recovery interventions that are desirable in the context of COVID-19 can also improve climate and disaster resilience or drive low-carbon development. When countries package these with interventions that address the inequalities and vulnerabilities exposed by the pandemic, they gain the opportunity to drive a sustainable economic recovery, reduce inequality, and build resilience. The note analyzes issues that the DMCs may face when designing recovery plans, and brings out the need to build climate and disaster resilience and low-carbon development into those plans. Potential characteristics that define a "good" recovery are identified and integrated into an initial assessment framework that decision-makers can use in assessing and prioritizing a package of suitable interventions. The note then considers how ADB could support its DMCs in implementing the interventions.

---

[3] Measures that support emission-intensive sectors or activities, would lead to an eventual increase in emissions, are potentially damaging to nature, or do not take climate and disaster risk into account.

II. Issues

A.   Recovery Planning Is an Opportunity to Promote Long-Term Transformation

5.   **Designing and financing the COVID-19 recovery is extremely challenging and will require a phased approach.** In the medium term, countries are expected to focus on economic stimulus measures to restart their economies and reengage their workforce, as well as on accompanying reforms, such as changes in monetary policy. Beyond this, countries must consider the impact of COVID-19 on their long-term development pathway and the steps they need to take, including transforming systems, institutions, and policies, to achieve long-term change based on a vision of a low-carbon and resilient future. Recovery plans must also be flexible enough to accommodate uncertainties around the future of the pandemic—the potential for future "waves" of the disease, the timeline for developing a vaccine, and the way in which governments, businesses, and consumers will continue to respond, among other things. Whether current restrictions will be maintained, loosened, or dispensed with, and how cautious firms or individuals will be in spending and engaging in activities, could alter the effectiveness of stimulus interventions.

6.   **Recovery planning must build the foundation for long-term prosperity.** The ability to develop and implement a COVID-19 recovery plan will vary between countries and should be viewed in the context of what experts predict could be a deep global recession. Low- and middle-income countries will have less fiscal space than high-income countries to go beyond their immediate emergency response. Debt levels, access to lending markets, and debt servicing costs will determine the capacity to implement fiscal measures to support the fiscal space available (EBRD 2020). Many countries, more exposed to exchange rate and maturity risks, and having lower credit ratings and shallower financial markets, are also likely to be less able or willing to increase the ratio of public debt to gross domestic product (GDP). Given the magnitude and uncertain duration of the COVID-19 impact a large number of countries will need substantial international support, in addition to emergency COVID-19 support they have already received, which will have implications for debt and fiscal positions (Bhattacharya and Rydge 2020). Existing resources may have to be redeployed to avoid very large deficits, which would cause a surge in national debt at a time when a sudden stop in capital inflows is a real risk. Governments will have to reexamine their budget revenues, expenditures and sources of financing, subsidies (including those for fossil fuels), and contingent liabilities (UNDP 2020a). The DMCs must think about all phases of the recovery and develop recovery plans that go beyond the immediate relief packages and strive to lift economies out of recession in the medium term, while laying a solid foundation for long-term prosperity. However, under great uncertainty and pressure, countries may find it challenging to consider long-term goals over short-term payoffs.

B.   Drivers for Integrating Climate and Resilience Considerations into the Recovery

7.   **Climate and disaster risks are rising and pose a serious threat to inclusive socioeconomic development.** Despite the scale of the COVID-19 crisis, as countries emerge from the initial emergency response phase, attention must return to the climate crisis and increasing disaster risk, and to the imperative of addressing that risk in parallel with the COVID-19 recovery. Countries do not have the time nor the financing to deal separately

with the COVID-19 and climate crises, and they must avoid letting concerns about the economic impact of the pandemic override the urgency of taking climate action. Climate change has resulted in prolonged heat waves, rising coastal sea levels, changes in rainfall patterns, and more intense and frequent extreme weather events. The lives and livelihoods of millions of people, especially the poor and vulnerable, are already affected, and progress in sustainable development is being undermined. Analysis done in 2019 estimated that climate change could push an additional 100 million people globally into poverty by 2030. By 2050, it could cut agricultural yield by up to 30%, and add more than $1 trillion each year to adaptation costs in urban areas (GCA 2019). GHG concentrations in the atmosphere are likely to keep rising, and temperatures never before experienced could be endured in the next few decades. Current emission reduction commitments made in nationally determined contributions (NDCs) are also insufficient, and would lead to a 3.2°C rise in temperature this century—well above the Paris Agreement target of 1.5°C (UNEP 2019). Further, because of rapid and unplanned development, with scant regard given to safety from natural hazards, more people and infrastructure are exposed to a wider range of disaster risks.

8.   **Prior to COVID-19 there was growing support for shifting to a more sustainable, inclusive, growth model that recognizes that climate action and economic growth can be complementary**. The pivotal Paris Agreement and Sustainable Development Goals of 2015, signaled a step up in global dedication to addressing climate change and achieving sustainable development. Accelerating the transition to clean and sustainable growth not only reduces the potential risk and impact of climate change, but provides an opportunity to harness the competitive advantages from more efficient resource use, preserving natural capital, and healthier more inclusive societies, among others. Analysis done by the Global Commission on the Economy and Climate in 2014 established that strong climate action by countries between 2018 and 2030 could, by 2030, generate over 65 million new low-carbon jobs, deliver at least $26 trillion in net global economic benefits, and avoid 700,000 premature deaths from air pollution. It also found that the $90 trillion infrastructure investment (an average of $6 trillion per year) required by 2030, for a high-carbon economy, across transport, energy, water systems and cities, would only need to be marginally increased by an estimated $270 billion per year for a low-carbon scenario. Furthermore, these higher capital costs could potentially be fully offset by lower operating costs, for example, from reduced expenditure on fuel. This cost is also small when considered in relation to the costs of inaction on climate change (GCEC 2014). This new agenda has broad-based support from governments, intergovernmental groups, and international institutions including Multilateral Development Banks (MDBs). For example, from the Group of Twenty (G20) leaders who affirmed their commitment to promoting strong, sustainable, balanced and inclusive growth (G20 2017), and the MDB's commitment in 2017, together with the International Development Finance Club, to align financial flows with the objectives of the Paris Agreement. Public and business support for climate action was also gathering significant momentum before the pandemic, and businesses and the finance sector have made important progress in incorporating climate-related risks in business planning.

9.   **Reductions in GHG emissions as a result of COVID-19 containment measures will not be sustained without transformative changes**. The restrictions governments have placed on mobility, and the drop in economic activity resulting from COVID-19, have reduced global GHG emissions in the short term. Estimates of the impact vary, because of uncertainty about the duration and extent of lockdowns, as well as the speed of recovery. The International Energy Agency (IEA) estimates that global GHG emissions could be around 8% lower in 2020 than their level in 2019—the largest annual drop since the Second World War, and six times larger than the reduction in 2009 due to the global financial crisis of 2008–2009 (IEA 2020a). According to the Climate Action Tracker, a 4%–11% reduction could occur in 2020, and between 1% above and 9% below 2019 emissions in 2021, depending on the magnitude and length of the economic turndown (CAT 2020). However, emissions are likely to rebound fairly quickly, and a short-term increase due to pent-up demand is also possible, based on experience from the global financial crisis. Global GHG emissions fell by 1.2% in 2009; however, the deployment of large stimulus packages was followed by a 5.9% rebound in 2010—well above the long-term average growth in emissions of around 2% (GCP 2011). The reductions in GHG emissions in 2020 have also been achieved through unsustainable methods, which severely restrict movement and economic activity. There has also been illness and tragic loss of life. While some behavioral changes that reduce GHG emissions, such as more flexible work arrangements with less travel, could stick, these will be nowhere near sufficient to achieve the emission reductions needed to address climate change. To limit the rise in global temperatures to 1.5°C above

pre-industrial levels, in line with the Paris Agreement, GHG emissions must drop by close to 8% every year from 2020 to 2030. This means that the emission reductions in 2020 would have to be repeated each year until the end of the decade (CAT 2020). The degree to which countries will be able to achieve this number or come close to it will depend greatly on how they choose to pursue their recovery. As modeling from the Climate Action Tracker shows, the rate and speed of the COVID-19 economic recovery is secondary to the speed and the degree to which economies transition from high- to low-carbon development (CAT 2020). Instead of focusing on "quick wins" that favor business-as-usual approaches and will lock in decades of high-carbon and unsustainable development, countries must look for opportunities to make transformative changes that promote a shift to a low-carbon growth trajectory, through investments and policy reform.

10.   **The pandemic has heightened risk perception, and could thus create momentum for investing in resilience**. COVID-19, by virtue of its immediate and localized impact, has triggered decisive government intervention, and a high degree of public compliance, worldwide. In contrast, climate and disaster risk has historically not been able to generate the same level of political or public support for action, as a large number of people feel spatially and temporally disconnected (consciously or unconsciously) from the threat. However, the lessons learned and the insights gained so far from the COVID-19 experience have increased awareness of all types of risks and built a strong case for making risk-informed decisions, including those required for low-probability, high-impact events. It has also led to broader recognition that risks across different sectors and themes need to be addressed holistically at a systems level, and that the immediate focus on reducing risks in the health sector must take into account that sector's links with other sectors, and the impacts on basic services, social protection, and food security, for example. This heightened awareness of risk may increase public support for action to address climate and disaster risk, as part of recovery and other plans, because of increased understanding of the importance of listening to science and using it as the basis for designing the response; the possibility that even low-probability events will occur, and their impact will escalate rapidly; the necessity of preparedness, and, hence, the need to invest in explicit resilience-building measures at all levels—local to global—to manage risks; and the fact that the preparation usually costs significantly less than the response (IMF 2020). This moment of increased awareness should be seized upon; COVID-19 recovery planning is an opportunity for governments to consider risks, including climate and disaster risk, recalibrate their priorities in the context of changing risks, and, where necessary, improve systems, raise standards, and pursue innovative solutions.

11.   **The scale-up of low-carbon and resilient investment will not compromise economic objectives**. A low-carbon, resilient recovery will not necessarily raise the required level of investment spending. Rather, it will require a shift in the nature of spending. Most countries have a wide variety of options for spending stimulus funds to create jobs and jump-start economic growth, but not all of these options will support a long-term vision of a low-carbon and resilient future. However, the list of possible recovery interventions that do promote this vision is long (see the Appendix) and includes few "new" ideas. Some interventions are already being implemented in a number of countries; others, after being assessed in the past, may have been turned down or deferred. However, new drivers in the current situation, including changed perceptions of risk and the importance of managing it, could boost the adoption of these measures, or affect the way in which they are implemented. The COVID-19 recovery could also present opportunities to speed up implementation of measures that have already been planned, because governments are looking to make stimulus investments quickly, and to improve the quality of investments by ensuring, for example, that countries invest in infrastructure that satisfies G20 Principles for Quality Infrastructure Investment (QII).[4] To make good decisions, countries need to understand clearly the long-term economic and social benefits they can derive from pursuing a low-carbon and climate-resilient recovery.

---

[4] The G20 Finance Ministers and Central Bank Governors endorsed the G20 Principles for Quality Infrastructure Investment at their meeting in Fukuoka, Japan in 2019, and emphasized the importance of value for money to guide infrastructure investment. This comprises a broader approach to sound project preparation practices over the project lifecycle, including the adoption of innovative technology, environmental and social sustainability, disaster and climate resilience, and good governance for procurement transparency and robust institutions. Investments aligned with these principles will help extend the life of infrastructure assets, increase the returns on the investment, and enhance social welfare.

12.   **There is a window of opportunity to transition toward low-carbon economies**. The increased momentum for climate action, combined with several factors that would not have existed during previous periods of economic difficulty, can be leveraged to support, or even increase, the benefits and the feasibility of a low-carbon resilient recovery.

- **Technology options for decarbonization have significantly improved**. Prices have also dropped, particularly for renewable energy technologies, battery storage, and electric vehicles. In the 2010–2019 period, the cost of power generation decreased by 82% for solar photovoltaics, by 47% for concentrated solar power, by 39% for onshore wind, and by 29% for offshore wind (IRENA 2020). The cost of battery storage has likewise been significantly reduced: the current costs of lithium ion batteries are only around 20% of their 2010 costs. In addition to utility scale generation and storage, the significant reduction in the costs of renewables and batteries offers opportunities to improve energy access and increase resilience in remote areas. But the drop in technology costs must be considered in conjunction with the impact of COVID-19 on the global energy system. Lockdowns and reduced economic activity have dramatically reduced energy consumption in 2020, contributing to extremely low and volatile fossil fuel prices and falling electricity prices. There is less incentive to be energy-efficient and to use renewable sources of energy. Combined with the ongoing economic uncertainty, these factors may put investment decisions at risk.[5] In this scenario, stimulus support for renewable energy will be an important part of a low-carbon and resilient recovery.

- **Current historically low oil prices can support subsidy reform**. Although low oil prices could make renewables less competitive, they also present an opportunity for subsidy and fuel taxation reform, including the removal of fossil fuel subsidies[6] and the introduction of carbon pricing (Dervis and Strauss 2020). Countries could consider subsidy "swaps" as well, with fossil fuel subsidies redirected to clean energy. The timing of such reforms could be politically sensitive, but they could also generate valuable budget revenue. As governments emerge from the crisis, many will be looking for measures to reduce debt and raise revenues (GCEC 2020). Numerous studies show that there are large economic benefits to be gained from reducing fossil fuel subsidies, with revenue gains estimated at 4% of global GDP (Coady et al. 2015), along with significant environmental and social benefits.

- **Growing climate ambition can be a starting point, and recovery planning does not have to start from scratch**. In line with their commitments under the Paris Agreement, countries were getting ready to submit new or updated NDCs in 2020. Countries have also prepared, or begun taking steps to prepare, long-term low-carbon climate-resilient strategies (LTS).[7] While the current pandemic has disrupted the process of updating NDCs in many countries, and may delay NDC submission, the obligation to submit the climate plans remains, and there is now an opportunity to align recovery plans with the national priorities defined in the NDCs, as well as long-term goals identified under LTS. Investment pipelines already identified in the NDCs can be taken forward, or accelerated, through the recovery process, thereby supporting long-term climate action aligned with national priorities. The same rationale applies to priorities set out in the existing national adaptation plans (NAPs), national disaster risk reduction (DRR) plans, and other related development plans.

- **There is private sector support for a sustainable economic recovery from COVID-19**. For example, a statement issued by international organizations across the energy, industry, finance, and civil society sectors, including BP, Shell, and Rio Tinto, calls for stimulus funds to be invested in "the economy of the future" (ETC 2020). The Investor Agenda, a global investor group representing members responsible for more than $55 trillion in assets, has warned governments to avoid focusing

---

[5]   Existing renewable energy projects are largely sheltered from lower electricity demand and falling electricity prices by fixed-price contracts and priority access to the grid (IEA 2020c).

[6]   The IEA costed fossil fuel subsidies in 2018 at more than $400 billion globally.

[7]   NDCs are submitted every 5 years. To enhance ambition over time, the Paris Agreement provides that each succeeding NDC will represent a progression over the previous NDC and reflect its highest possible ambition. Countries agreed that by 2020, those countries with an NDC time frame up to 2025 would communicate a new NDC, and those with an NDC time frame up to 2030 would communicate or update an existing NDC (Decision 1/CP.21). A decision made at the 24th Conference of Parties (COP24) in Katowice, Poland (Decision 1/CP.24), in 2018 explicitly reiterated this request to revise NDCs by 2020. Also under the Paris Agreement, countries are encouraged to formulate and communicate Long Term Strategies, taking into account the national development context.

on short-term, high-emitting projects, stating that "as governments pursue efforts to recover from this economic downturn, they should not lose sight of the climate crisis. They must factor in the foreseeable, acute, systemic and compounding climate-related economic and financial risks" (The Investor Agenda 2020).

13.   **Investments must be accompanied with policy change**. For the recovery measures to be sustainable, countries must implement supporting policies that provide the right long-term economic incentives and market signals to sustain the changes in behavior or systems beyond the period where the stimulus will be available. The processes for policymaking need to be upgraded to better manage complexity, deploy open policy-making approaches and apply systems and design thinking to strengthen impact and integration, while expanding government's ability to plan for a range of possible threats and stresses (ADB 2020c). Countries will also need to manage the structural changes required for transitioning toward low-carbon economies, including addressing the resulting distributional impact,[8] and supporting a just transition that promotes economic diversification and inclusion.[9] In addition, countries may have to dismantle existing policies that discourage low-carbon and resilient development, and, in their place:

- introduce economic incentives for low-carbon products or sectors, and remove existing disincentives (e.g., import duties on solar panels);

- make policy changes, such as adopting carbon pricing or tax regimes, or electric vehicle policies;

- introduce new standards or regulations, such as climate and disaster-proofing standards for infrastructure;

- mainstream climate and disaster risk into national development planning and budgeting; and

- provide strategic support for research and development (e.g., development of climate-resilient crops) and pilot projects (e.g., green infrastructure for flood risk management).

Experience from the global financial crisis supports the benefits of "green" stimulus but strongly confirms the need for policy change, for the stimulus funding to be effective over the long term. There is evidence from the 2008–2009 economic recovery that well-targeted stimulus interventions, including investments in renewable energy, energy efficiency, and public transport and natural (green) infrastructure, generated more jobs than traditional alternatives, and delivered high fiscal multipliers. In 2008–2009, around 16% of the global stimulus package (of around $3.3 trillion) was spent on "green" measures, mainly in countries within the G20. The Republic of Korea had one of the highest proportions of "green" spending, estimated at around 79% ($59 billion) of its total stimulus package. By the third quarter of 2009, it had one of the highest growth rates in the Organisation for Economic Co-operation and Development (OECD) (Bhattacharya and Rydge 2020). The People's Republic of China indicated that around a third of its stimulus spending, ($219 billion) was "green"; the United States (US) green spending was around 12% ($118 billon); and up to 60% ($23 billion) of the European Union's collective stimulus was similarly directed at green initiatives (Jacobs 2012). Around half a million net jobs were estimated to have been created by the environmental elements of the US stimulus package, with as many as 960,000 by the similar measures taken in the Republic of Korea (Jacobs 2012). However, most countries did not introduce market reforms, develop complementary policies, or remove prevailing economic incentives that encouraged emission generation and environmental degradation, such as fossil fuel subsidies. Neither did they introduce carbon and environmental taxation or other pollution abatement measures. As a result, the stimulus packages did not have a lasting impact after they were withdrawn (Mission 2020). Additionally, as the proportion of spending on green measures varied widely between countries, effectiveness in "greening" the global economy was limited (Barbier 2010). The IEA recently considered these issues and examined lessons from the global financial crisis that could be applied to

---

[8]   For example, by introducing carbon pricing together with social protection, such as using carbon revenue to compensate low-income households.

[9]   The Paris Agreement requires Parties to take into account "the imperatives of a just transition." The term is variously defined, but essentially means that development pathways, including decarbonization, ensure environmental sustainability, as well as decent work, social inclusion, and poverty eradication. At COP24, over 50 heads of state and governments signed the Silesia Declaration on Solidarity and Just Transition, which reaffirmed this commitment to just transition, and the relationship between responses to climate change and equitable sustainable development and poverty eradication. A just transition that 'leaves no one behind' is also directly linked to SDGs such as SDG7 (Affordable and Clean Energy), SDG8 (Decent Work and Economic Growth), SDG10 (Reduced Inequalities), and SDG13 (Climate Action).

the COVID-19 recovery. According to the IEA analysis, feed-in tarrifs amounting to $93 billion were the most important policies in European Union countries for mobilizing investments in solar photovoltaic and wind energy (IEA 2020b). Most of these were not part of the green stimulus in a narrow budgetary accounting sense but were introduced at the same time, giving the energy system a major boost. The IEA report also concluded that stimulus funding is most successful where funding programs are based on proven policy schemes; are targeted at technologies that are ready for deployment; consider the wider benefits, including social, energy security, and industrial policy objectives; and tackle structural barriers to investment.

14.   **Strengthening institutional arrangements and good governance will also be key to a sustainable recovery** particularly as many countries will have significantly larger public debt, and limited fiscal headroom for many years after the pandemic and associated stimulus spending. Sound public financial management will be important to ensure public expenditure is prioritized, accountable and efficiently delivered. This could be challenging in an environment where decisions on stimulus spending are being made relatively quickly and under pressure.

C.   Resilient and Low-Carbon Interventions in Support of Economic Recovery

15.   **A package of recovery interventions with a range of characteristics is needed.** Different countries will have different criteria for recovery interventions, depending on many factors such as a country's main economic sectors, COVID-19 experience, recovery budget, and access to international support. However, as suggested in the literature analyzed in this technical note, interventions that will enable a "good" economic recovery from COVID-19 share common characteristics with economic stimulus initiatives that were successfully launched in the past. These interventions

- are usually designed to generate jobs and stimulate economic activity, and may involve actions with a short implementation timeline, so that the stimulus funds can be quickly deployed and people can get back to work in a timely manner (and possibly with less need for safety nets);

- are labor-intensive, particularly in the early stages;

- promote skills development, with objectives such as re-skilling unemployed workers from sectors that have been heavily affected by COVID-19, or preparing for a low-carbon future;

- have a strong supply chain, which could include considerations such as more localized or more diversified sourcing; and

- have high economic multipliers.

For a low-carbon, resilient recovery, these criteria still apply, but need to be supplemented with other essential features that support a "build back better" approach to stabilizing the economy, building resilience, and improving the quality of life in the long term (IMF 2020; Jotzo, Longden, and Anjum 2020; OECD 2020; Green Stimulus Proposal 2020). Countries are likely to a have a variety of options for public spending and investment that can create jobs and have high economic multipliers, but these short-term requirements must be assessed in parallel with long-term outcomes centered on building resilience, decarbonizing economies, and delivering on the SDGs, including

- measures that contribute to the productive asset base, e.g., by providing climate- and disaster-resilient infrastructure, or by improving digital communications;

- promote long-term transformation, e.g., by supporting innovation or re-skilling; and

- have positive social and environmental outcomes.

COVID-19 has led to asymmetric socioeconomic impact across sectors, within and between countries. Therefore, where possible, recovery interventions should be directed at affected groups or regions, foster equitable growth and investments, and provide fairer access to employment opportunities. This is particularly critical in addressing the gender-related implications for women and girls, particularly those from vulnerable groups. Different countries may adopt more nuanced requirements, or define different criteria for different phases of recovery or for different regions or sectors (e.g., specific environmental objectives such as reducing air pollution or improving biodiversity), but if a clear vision has been set for a recovery leading to a low-carbon, climate- and disaster-resilient future, it will help to ensure that all of the defined criteria support this unified approach. In view of the scale of the current crises, and the various phases of recovery, countries will need to implement a package of recovery interventions that collectively cover all of the desired characteristics. Besides providing the required stimulus, the intervention package must also address the underlying barriers to ensure that the intended changes are sustained.

16.   **Low-carbon and resilient interventions can support a sustainable economic recovery, address vulnerabilities, and build resilience.** A number of low-carbon and resilient recovery interventions, including direct investment as well as policy reform and capacity building (hard and soft measures), have been identified (see the Appendix). Analysis shows that many of these interventions have the abovementioned characteristics underpinning a "good recovery". Or

- **Clean energy investment performs well against the characteristics of a "good recovery," as well as delivering climate and development benefits.** For example, building insulation retrofits and clean energy infrastructure are labor-intensive investments in their early stages, and can deliver high multipliers and have high returns over the longer term by driving down the cost of the clean energy transition. One widely cited model suggests that every $1 million in spending generates 7.49 full-time jobs in renewable energy infrastructure and 7.72 full-time jobs in energy efficiency, but only 2.65 full-time jobs in fossil fuels (Garrett-Peltier 2017). Strengthening women's roles and employment in renewable energy can also contribute to the achievement of several SDGs.

- **Green investments have high economic multipliers.** According to recent estimates made by the Climate Council of Australia (as part of its proposed Clean Jobs Plan), every A$1 invested by the government in pilot-scale hydrogen facilities unlocks A$4 in social value; utility-scale renewable energy returns A$3 of every dollar invested; and electric vehicle infrastructure, improved organic waste collection and processing, and community-scale energy and storage returns A$2 for every dollar invested (Climate Council and AlphaBeta Australia 2020). Investment in restoration can give rise to a variety of new income streams, which could boost smallholder farmers' incomes, and restoring a mere 12% of the world's degraded agricultural land could feed 200 million people by 2030, while also strengthening climate resilience and reducing emissions (GCEC 2014). Green investment offers opportunities to increase the proportion of green jobs held by women and improve their participation in the workforce.

- **Investment in low-carbon infrastructure yields dividends.** Because of technological advances and the long-term reduction in the cost of renewable energy and other clean technologies, the benefits could exceed those obtained in previous economic recoveries. Many countries, enjoying the unintended benefit of reduced air pollution amid this pandemic, may be willing to give priority to maintaining clean air in their economic recovery plans. Other measures that support the decarbonization of the electricity system, such as distributed generation and storage and a more flexible power system, can also improve resilience to future shocks, including climate and disaster shocks. Better access to electricity, through off-grid or mini-grid renewable systems, can have many benefits as well for employment, well-being, health, and societal resilience, besides enhancing affordability, reducing the environmental footprint, and cutting network infrastructure costs.

- **Greater use of digital technologies can promote decarbonization and increase resilience while also supporting recovery.** An example would be the increased use of digital technologies for supply chain management to improve resilience, reduce the risk of supply chain disruptions, and address vulnerabilities exposed by COVID-19, while at the same time improving resource and energy efficiency, with potential reductions in GHG emissions.

- **Preserving and investing in natural capital can address medium-term needs while laying the groundwork for long-term change.** Protecting natural capital, such as mangroves in coastal areas, to lessen the impact of storm surges, and conserving water retention ponds to reduce flooding, brings natural features into the built landscape for environmental sustainability. Such interventions can be funded through public works programs as part of active labor market programs, which are likely to be prevalent in the recovery context. These programs, possibly headed by local authorities in consultation with the community, can take sustainable approaches to local employment generation that are community-owned and community-driven. For example, evidence shows that protecting mangroves in coastal areas of Viet Nam has various benefits, including keeping low-lying areas safe from storm surges with the help of dikes, growing forests, and generating additional income for coastal communities through aquaculture products (IFRC 2010). Investing in natural capital for resilient recovery can also have poverty reduction, enhanced development, and other welfare co-benefits, each of which is critical to the COVID-19 recovery. Women also have a key role in biodiversity conservation, and in the promotion of natural capital.

- **Investment in resilient infrastructure is cost-efficient and has a high economic multiplier.** Public spending on infrastructure is a common feature of economic stimulus packages, and is likely to be an area of significant investment in the COVID-19 recovery in many countries. While the incremental costs of making infrastructure resilient to future shocks and stresses may vary depending on project location, scale, and complexity, these costs are largely found to improve the profitability of the investments. Recent estimates suggest that investing in more resilient infrastructure in low- and middle-income countries produces a net benefit of $4 for each dollar invested. Climate change makes action on resilience even more necessary and attractive: on average, the net benefits of taking resilience measures are doubled. However, spending on infrastructure will only deliver these benefits if the barriers to delivering resilient infrastructure are addressed, including making project preparation efficient and effective, and ensuring the focus is on "spending better" (Hallegatte, Rentschler, and Rozenberg 2019). The IMF reviewed public investment management frameworks of 30 low- and middle-income countries, and it concluded that 30% of potential economic benefits of public investments are lost due to inefficiencies in the public investment process (IMF 2018). Governments need support to prepare high-quality projects in line with the G20 QII Principles to ensure efficiency, affordability and sustainability.

- **Climate- and disaster-resilient agriculture value chains can provide employment and bring about transformational change.** The pandemic has brought to light the importance of resilient value chains in agriculture. In many of the DMCs, the sector provides most of the needs of a large proportion of the population and employs more female workers than any other sector. Investing in value chains that are more resilient to different types of shocks and stresses, including climate- and disaster-related shocks, will be critical to the recovery process. An important aspect of resilience-building activities will be promoting localized supply chains, thus helping to reduce GHG emissions, strengthening the local economy, and directly contributing to the well-being of the local population. Investments in resilient supply chains must look beyond production and include a range of risk management measures to inform inputs, processing, marketing, and consumption. They should also build awareness of sustainable agricultural practices among farmers and make them more capable of transitioning to such practices. Targeted and gender-responsive interventions should be carried out in both rural and urban areas, given their different vulnerabilities.

- **Improved preparedness and early warning systems produce long-term benefits.** COVID-19 has likewise drawn attention to the need to improve preparedness to deal with shocks related to health or other emergencies. Investments in improved local preparedness and in end-to-end early warning systems can generate significant benefits and can be transformative. For example, investments in both physical and institutional resilience, including forecasting and early warning capabilities, enabled the evacuation of around 1 million people from low-lying areas to cyclone shelters, schools, and other safe buildings before Tropical Cyclone Fani hit the state of Odisha in India, in May 2019. Consequently, only 89 people died, compared to the 10,000 deaths that occurred when the same state was battered by a super cyclone 20 years earlier. Recent estimates show that investing $1.8 trillion globally in five climate adaptation areas, including strengthening early warning systems in 2020–2030, could generate a total of $7.1 trillion in new benefits (GCA 2019). Future investments to improve preparedness should consider the compound impact of different hazards and crises, similar to the devastation wrought by Tropical Cyclone Amphan in India and Bangladesh in May 2020, amid the COVID-19 pandemic.

- **Investments in adaptive social protection can build longer-term systems to protect the poor and vulnerable population from future crises.** Investments to integrate adaptive and shock-responsive features into the design of social assistance programs can help build the adaptive capacity of the poor and vulnerable, including women, to deal with future crises. Some examples are as follows: using climate risk information to improve the targeting of social assistance programs; introducing flexible design features that allow cash transfer programs to be scaled up after a climate shock to support more beneficiaries, or to increase support provided to existing beneficiaries; linking cash transfer programs to early warning systems so that the shock-responsive features can kick in before disaster strikes; and strengthening links between social assistance programs and sustainable livelihood programs to build resilience.

17. **Five key intervention areas for recovery have been identified.** The potential for a low-carbon and resilient recovery to deliver strong economic benefits was supported by the findings of a study completed in May 2020 (Hepburn, O'Callaghan, Stern, Stiglitz, and Zenghelis 2020), involving a survey of 231 central bank officials, finance ministry officials, and other economic experts. It assessed the performance of 25 major fiscal recovery archetypes across four dimensions—speed of implementation, economic multiplier, climate impact potential, and overall desirability—and recommended five key intervention areas for recovery

- investment in clean physical infrastructure in the form of renewable energy assets, storage (including hydrogen storage), grid modernization, and carbon capture and storage (CCS) technology;

- building efficiency spending for renovations and retrofits, including improved insulation, heating, and domestic energy storage systems;

- investment in education and training, to address the immediate problem of unemployment due to COVID-19 and structural shifts resulting from decarbonization;

- natural capital investment for ecosystem resilience and regeneration, including restoration of carbon-rich habitats and climate-friendly agriculture; and

- clean research and development (R&D) spending.

In addition, for low- and middle-income countries, rural support scheme spending, particularly associated with sustainable agriculture, ecosystem regeneration, or acceleration of clean energy installations, was recommended (Hepburn, O'Callaghan, Stern, Stiglitz, and Zenghelis 2020).

III. Opportunities for a Low-Carbon and Resilient Recovery

A.   An Assessment Framework for Climate and Resilience Benefits

18.   **ADB has developed an initial assessment framework** (Table) to assist its DMCs in evaluating interventions for a low-carbon and climate- and disaster-resilient recovery, such as those found in the Appendix. The framework provides a structured process and visual aid for evaluating, and comparing, the potential of the interventions to achieve the intended recovery objectives by assessing the interventions against a set of key requirements for COVID-19 recovery—or the characteristics of a "good recovery," as discussed in section IIC. Decision-makers can use the framework, tailored to reflect each DMC's specific circumstances and recovery objectives (as discussed in para. 16), to select and prioritize a package of interventions that will collectively achieve those objectives and promote climate resilience through their medium-term recovery and longer-term transformation efforts. The framework can also help decision-makers understand the potential negative implications of certain interventions. A rapid initial assessment of potential interventions based on a qualitative assessment of measures can be made (as shown in the table), or the framework can be developed further to allow deeper analysis, including, for example, a separate assessment for each phase of recovery or a quantitative assessment of the performance of measures. Briefly, the steps involved in this assessment would include the following:

- **Step 1: Define a clear vision for a recovery that leads to a climate- and disaster-resilient future.** A clear vision will build confidence, ensure a unified approach to the recovery, and allow the definition of medium-term recovery and long-term transformation objectives. With this vision, countries can define principles to guide the recovery, such as the following:

  - » putting people and their health first, to ensure that no one is left behind;

  - » taking a "build back better" approach to stabilizing the economy, promoting equitable growth and investments that benefit all, and strengthening the localization of supply chains;

  - » promoting a transformational shift to a low-carbon and resilient development pathway (and long-term "net zero") and supporting a just transition, where benefits are shared equally;

  - » using the COVID-19 recovery to accelerate delivery of the SDGs;

  - » supporting investments that contribute to the productive asset base for the future; and

  - » committing to policy reforms, institutional change, and capacity building to sustain the results of the recovery.

- **Step 2: Define the objectives, requirements, and "good recovery" criteria.** These will reflect the country's short-term economic recovery criteria, such as job creation, speed of implementation (e.g., "shovel-ready," or within 6 months), strong supply chain (e.g., shorter, more diversified, more localized), and long-term requirements for achieving a sustainable low-carbon and resilient recovery, including contribution to the productive asset base. The assessment can also be disaggregated at a lower level. For example, criteria could be defined for specific phases, regions, or sectors.

- **Step 3: Define how the performance of each measure against the "good recovery" criteria will**

be assessed. In the example presented in the table, a qualitative "high"–"medium"–"low" assessment was used. This approach supports a more rapid assessment. A more detailed analysis could be done—for example, by quantifying the "high"–"medium"–"low" assessment or assigning weights or scores to certain criteria.

- **Step 4: Define the criteria for assessing climate and resilience benefits.** These need to reflect national priorities, including those in NDCs, NAPs, etc., and consider long-term decarbonization (including net zero GHG emissions), as well as strengthening resilience and addressing vulnerabilities.

- **Step 5: Draw up a "long list" of potential low-carbon and resilient recovery interventions to be assessed.** This includes reviewing relevant national plans, such as NDCs, NAPs, National DRR Plans and other related development plans, as well as associated investment pipelines, plans or programs that have been identified. Organize these into relevant groups (e.g., direct investment, policy and regulatory changes, institutional strengthening, and capacity development).

- **Step 6: Prepare the assessment framework.** Reflect the decisions made in Steps 1–5 in the framework, and identify and prioritize the intervention measures to be included in the package of recovery interventions.

- **Step 7: Assess financing and supporting measures.** Review the proposed package of interventions considering the following:

  » **Conditionality stipulations**, which may be required in cases where "brown" recovery interventions are necessary for providing short-term relief to key sectors, or to sectors with an important role in decarbonization. These stipulations could include any of the following: pledges to reduce emissions, or plans for net zero; measures implemented to ensure that funds used support workers and the creation of good-quality jobs, or skills development for a low-carbon and resilient future; pledges to use funds to build more resilient and lower-carbon supply chains; and use of funds to increase preparedness for extreme weather events. These could be further supported by the focus of central banks on making sure investments—under the Network of Central Banks and Supervisors for Greening the Financial System (NGFS, n.d.)—consider transition risks associated with future climate policy or taxes (thus avoiding the risk of stranded assets).

  » **Potential sources of financing** for the measures to confirm feasibility and identify specific climate finance–related opportunities. Among the options to be considered are raising domestic revenue, mobilizing private sector finance (including public–private partnerships), bringing in innovative and "green" financial products (such as green or climate bonds), and leveraging international climate finance. These need to take into account, and align with, the country's detailed process for developing its recovery budget.

  » **Supporting policy and institutional changes** that will be necessary for long-term transformation and the sustainability of recovery measures, as discussed in para. 11.

As part of the assessment process, the DMCs could also develop a checklist for assessing the alignment of recovery interventions with international agreements, such as the Paris Agreement and the Sendai Framework, incorporating aspects such as the following: (i) allowing fossil fuel investments only if they are clearly part of a low-carbon transition; (ii) including carbon prices in the economic appraisal of support interventions, to ensure that stimulus activities are appropriately priced; (iii) checking whether the proposed actions align with a DMC's NDC, NAP, and national DRR plan; and (iv) including a climate and disaster risk assessment check that reflects issues related to gendered vulnerabilities.

Table: Framework for Assessing Low-Carbon and Resilient COVID-19 Recovery Interventions

Recovery Measures	Climate and Resilience Results and Benefits	Type of Measure	Requirements of COVID-19 Recovery Measures							
			Short Implementation Timeline	High Employment Intensity	Skills Development	Strong Supply Chain	High Economic Multiplier	Contribution to the Productive Asset Base	Support for Long-Term Transformation	Positive Environmental and Social Outcomes
Low-Carbon Development										
Investment in low-carbon (renewable) energy production and energy storage infrastructure		DI	Medium	High	High	Medium	High	High	High	High
Extension and modernization of the grid to support higher renewable penetration		DI	Medium	High	High	Medium	High	High	High	High
Public procurement program for the purchase and installation of energy-efficient appliances, lighting, and digital devices for public buildings		DI	High	High	High	High	Medium	High	High	High
Incentives for home renovations and retrofits, including low- and zero-energy measures, in affected regions		P&R	High	High	High	Medium	Medium	Medium	High	High
Introduction of green tax regimes, e.g., carbon taxes, carbon price floor (for industry)		P&R	Low	Low	High	Medium	High	Medium	High	High
Planning of urban green redevelopment/regeneration and sustainable spaces (smart cities)		T	Low	Low	High	Medium	High	High	High	High
Development and scale-up of radical transport (universal and comprehensive public transport/car-free movement)		T	Low	Low	High	Medium	High	High	High	High
Climate and Disaster Resilience										
Reorientation of labor market programs to support resilience-building measures (e.g., water resource conservation, reforestation)		DI	High	High	Medium	Low	Medium	Medium	High	High
Development of climate-resilient agricultural value chains		DI	Medium	High	High	High	High	High	High	High
Investment in protective infrastructure to strengthen resilience (e.g., coastal protection, flood defense)		DI	Medium	Medium	Medium	Medium	Medium	Medium	High	High
Active labor market policies and economic stimulus to support job creation in resilience sectors		P&R	Medium	High	High	Medium	High	Medium	High	High
Introduction of policy reforms to enhance resilience (e.g. payment for ecosystem service schemes)		P&R	Medium	Medium	Low	Medium	Medium	High	High	High
Transformation of rural food and land-use systems, including a shift to sustainable and resilient production		T	Low	Medium	High	High	High	High	High	High
Risk-sensitive land-use management		T	High	Low	High	Medium	Medium	Medium	High	High

Note: Assessment will vary by country, or even by region. Assessment is for illustrative purposes only.
Source: ADB 2020.

Climate and Resilience Results and Benefits

- Addresses vulnerabilities
- Targets COVID-19 affected sectors or populations
- Targets disadvantged groups (e.g., regional, women)
- Builds long-term resilience
- Supports development of high level technology (e.g., low-carbon)
- Supports long-term decarbonization
- Consistent with national policies and plans

Type of Measure

DI = direct investment

P&R = policy and regulatory

T = transformative

Potential to Achieve Recovery Objectives

- Low
- Medium
- High

B.   Opportunities for ADB to Support its DMCs

19.   **ADB can play a critical role in supporting its DMCs to pursue low-carbon and resilient development through recovery-related planning and implementation**, thus maintaining its strong leadership in climate change and disaster risk management in line with Strategy 2030 (ADB 2018) and ADB's climate targets. Each of the operational priorities of Strategy 2030 and the call to "build back better" in the context of the COVID-19 recovery are strongly linked (Figure 1).

Figure 1: Illustrative Examples of Low-Carbon and Resilient Recovery Interventions Align with Strategy 2030

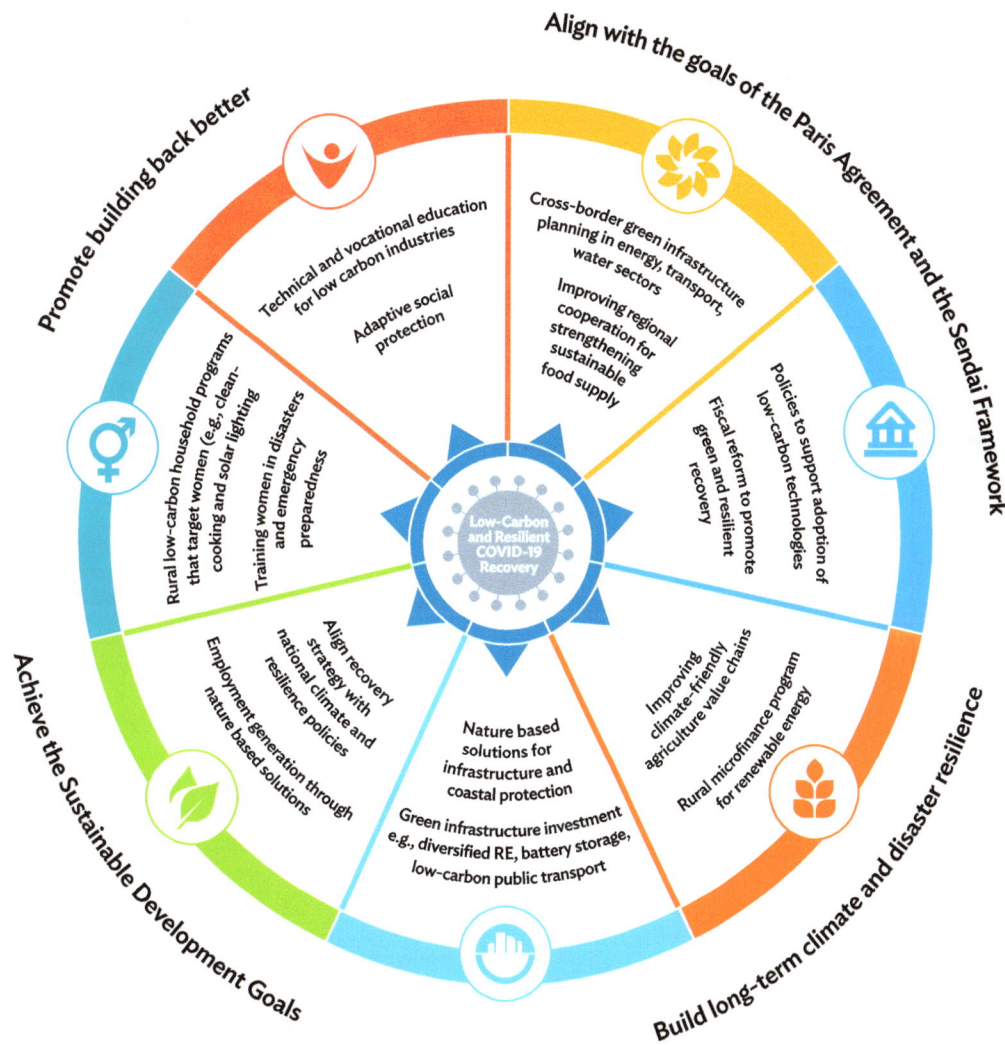

Source: ADB 2020.

Legend for Operational Priorities (OP) of S2030

OP1: Addressing Remaining Poverty and Reducing Inequalities

OP2: Accelerating progress in gender equality

OP3: Tackling Climate Change, Building Climate and Disaster Resilience, and Enhancing Environmental Sustainability

OP4: Making cities more livable

OP5: Promoting Rural Development and Food Security

OP6: Strengthening Governance and Institutional Capacity

OP7: Fostering Regional Cooperation and Integration

20.   **Operational priority 3 of Strategy 2030 is tackling climate change, building climate and disaster resilience, and enhancing environmental sustainability.** Key responses identified under operational priority 3 are wholly relevant to a low-carbon and resilient COVID-19 recovery - including scaling up support to address climate change, disaster risks, and environmental degradation; accelerating low GHG emission development; ensuring a comprehensive approach to build climate and disaster resilience; ensuring environmental sustainability; and increasing focus on the water–food–energy nexus (Figure 2), meaning ADB support to its DMCs to adopt a low-carbon and resilient recovery will align with ADB's exisiting efforts towards operationalizing its strategic framework, as well as aligning operations with the goals of the Paris Agreement (Figure 3). It will also support ADB in achieving its climate finance targets, ensuring that at least 75% of its projects focus on climate change mitigation and adaptation, while providing $80 billion in climate finance cumulatively by 2030. While the scope or implementation of ADB's support may be different during the COVID-19 recovery, and as a result of the impact of COVID-19, the Paris Alignment principles adopted by ADB, together with other MDBs, are still essential to guide ADB's activities towards supporting the goals of the Paris Agreement. Governments in many countries have limited capacity to deal with COVID-19. Their revenues having been curtailed by the crisis, they have less fiscal space for the necessary reforms and investment. ADB must continue to support its DMCs to prioritize climate and resilience action, despite the challenges they face as a result of COVID-19, through technical assistance, help in planning institutional and policy reforms and setting these in motion, financing, and integration of these considerations into country partnership strategies.

21.   **ADB could offer targeted support** to promote low-carbon development and climate and disaster resilience more explicitly by

- Providing technical assistance to help the DMCs with recovery planning, including support in using the proposed assessment framework (described in section IIIA) to identify and prioritize investments in climate and disaster resilience and low-carbon development in recovery packages. An analysis of possible policy and regulatory reforms, as well as stimulus activities, could be part of the support provided. Where appropriate, ADB could access resources for technical assistance from its **Climate Change Fund** to integrate climate and disaster resilience considerations in recovery planning processes.

- Supporting the reassessment of the DMCs' existing pipeline of projects or activities, to help align stimulus packages with NDCs, NAPs, national DRR plans, and other relevant plans, and identify priority projects on the basis of low-carbon and resilience criteria (focusing on projects that are ready or close to ready). Where NDCs do not yet include concrete pipelines, ADB can support its DMCs in developing the pipelines as part of their climate investment plans. This support could be provided under ADB's existing technical assistance platform, **NDC Advance**, which is helping the DMCs to meet their climate objectives by translating their NDCs into climate investment plans, facilitating better access to external public and private climate finance, and developing methods and tools for monitoring progress on climate action.

- Supporting the DMCs to align COVID-19 recovery plans with the development of long-term low GHG and resilient development strategies, or LTS, under the Paris Agreement, including Just Transition elements. LTS are essential for the transition to net zero emissions and resilient economies, and considering long-term goals and development pathways can help guide the DMCs with short-term decision-making for COVID-19 recovery.

Figure 2: Strategy 2030 Operational Priority 3

Operational Priority 3

Tackling climate change, building climate and disaster resilience, and enhancing environmental sustainability

- Integrated approach in country partnership strategy/country operations business plan
- Deploy approaches for capturing co-benefits in coordination with other operational priorities
- Promote innovative clean technology
- Expand private sector operations
- Build partnerships with think tanks, nongovernment organizations, academe, and private sector
- Access to finance: use of concessional finance in a targeted and catalytic way–maximizing delivery of outcome

Targets: 75% committed operations (3-year rolling average) and **$80 billion** of own resources (2019–2030, cumulative) will support climate actions

Strategic Operational Priorities	Operational Approaches	Sub-pillars
1 **Mitigation of climate change increased**	Clean energy *Green business and jobs* Sustainable transport and urban development *Clean air and water, waste management*	• Access to climate finance increased • Capacity of developing member countries to implement climate actions enhanced • Low-carbon infrastructure improved • Renewable energy capacity increased • Low-carbon development solutions implemented
2 **Climate and disaster resilience built**	Climate-smart agriculture and sustainable land use Climate and disaster Resilience *Physical (climate-proof), eco-based, financial, social, and institutional*	• Integrated flood risk management measures supported • Resilience building initiatives implemented • Finance preparedness for post-disaster response enhanced • Planning for climate change adaptation and disaster risk management improved • Infrastructure assets made more resilient
3 **Environmental sustainability enhanced**	Water–food–energy security nexus Air and water pollution management Natural capital and healthy oceans *Environmental governance*	• Pollution control infrastructure assets implemented • Pollution control and resource efficiency solutions promoted and implemented • Conservation, restoration and enhancement of terrestrial, coastal and marine areas implemented • Solutions to conserve, restore, and/or enhance terrestrial, coastal, and marine areas promoted and implemented • Water–food–energy security nexus addressed

Source: ADB 2019.

- Identifying opportunities to bring forward relevant projects that are already in the ADB pipeline, or were recently proposed, and speeding up the appraisal and approval process, without compromising due diligence. This could include examining opportunities under the **Asian Development Fund 13** thematic pool on scaling up support for disaster risk reduction and climate change adaptation in group A and group B DMCs to support resilience building in sectors that have been exposed as being vulnerable in the current pandemic.

- Examining opportunities under relevant existing projects to support COVID-19 recovery-related activities, such as the proposed **Community Resilience Partnership Program**, which is currently being designed to support the DMCs in scaling up investments in local resilience that explicitly address the nexus between poverty, gender, and climate and disaster risk, and the **Urban Platform for Climate Finance**, which will assist cities in implementing climate-resilient and low-carbon investment plans.

- Helping the DMCs to gain access to opportunities for blended finance, recognizing that multi-donor cofinancing, including access to resources from multilateral climate funds such as the Climate Investment Funds, the Global Environment Facility, and the Green Climate Fund, will be critical to achieving a low-carbon and resilient recovery. This assistance could include, where possible, supporting the DMCs in mobilizing financing quickly, for example, through available COVID-19 financing platforms and resources.

- Working through the ADB managed **ASEAN Catalytic Green Finance Facility** which is helping the DMCs to incorporate green finance approaches and mechanisms into their recovery strategies including through the preparation of country-specific green finance recovery proposals.

- Supporting the DMCs in obtaining access to carbon finance through new market mechanisms under Article 6 of the Paris Agreement where appropriate, including providing technical, capacity building, and policy development support through the **Article 6 Support Facility**, as well as mobilizing carbon finance through ADB's ongoing **Japan Fund for the Joint Crediting Mechanism and the Climate Action Catalyst Fund**, which is being designed.

- Continuing to support policy and regulatory reform, such as the introduction of carbon pricing and other environmental taxes, and the removal of fossil fuel subsidies to allow countries to find fiscal space through reforms and budget reallocations.

- Supporting the review of budgetary priorities and the adoption of framework evaluation for investment toward low-carbon and resilience (extending existing climate-related financial expenditure and climate budget tagging work). This would also cover support for climate and disaster risk mainstreaming into national development planning and budgeting.

Figure 3: ADB and Paris Alignment

Six building blocks and principles jointly agreed by the MDBs as core areas for alignment with the Paris Agreement

Methods and tools identified are screened based on criteria:

- Rigor
- Potential for cross MDB standardization
- Applicability
- Ease of implementation
- Feasibility

*Implemented according to ambition and timeline that can be differentiated by each bank

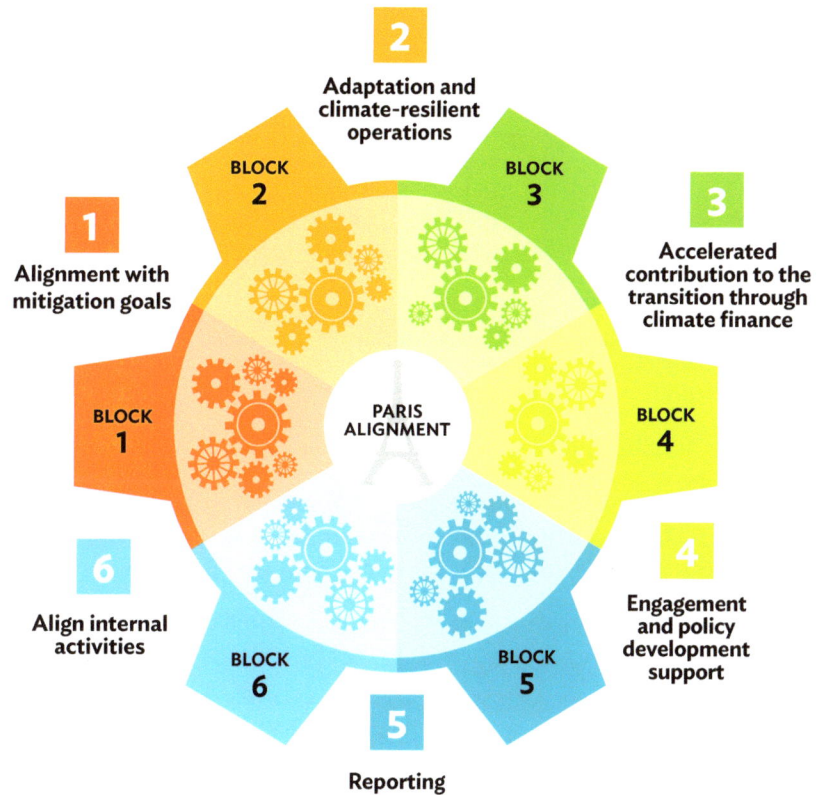

ADB = Asian Development Bank; MDB = multilateral development banks.
Source: ADB 2019.

22. **A second stage following from this technical note is envisaged**, to

- develop further analyses and case studies on how to build climate and disaster resilience, and low-carbon development elements, into sector- and country-specific recovery plans;

- examine the ADB pipeline and work programs to identify examples of the types of interventions identified in this technical note that could form part of the recovery support, and from these develop case studies on how to incorporate climate and disaster resilience, and low-carbon development elements, into the recovery plans; and

- provide suggestions for the further development of the joint multilateral development bank framework for alignment with the objectives of the Paris Agreement, in the light of the COVID-19 recovery interventions.

APPENDIX: Climate- and Disaster-Resilient and Low-Carbon Recovery Interventions

The relevant interventions proposed in the literature reviewed in this technical note were grouped into those that primarily address climate and disaster resilience, and those that are focused on low-carbon development, while recognizing that some interventions have the potential to achieve both. The potential recovery interventions were classified as:

- **direct investment**, e.g., in low-carbon or resilient infrastructure;

- **policy and regulatory reform**, e.g., tax reform, regulatory standards; or

- **transformational change**, i.e., deeper, more fundamental shifts to existing systems.

Countries also need to raise **finance**, e.g., develop new financial instruments and structures, and undertake **capacity building and institutional strengthening** to support long-term change and a sustainable recovery. Potential options for doing this are presented here after the recovery interventions.

Some interventions could be targeted to respond specifically to the health crisis and exposed vulnerabilities, whereas others are more general and may not be new, but have high relevance to the coronavirus disease of 2019 (COVID-19) recovery.

A. Climate and Disaster Resilience

Direct Investment	Investment in hard (engineered) resilience infrastructure (e.g., coastal protection) as part of a stimulus package (e.g., to create jobs, local investment)
	Investment in nature-based solutions (e.g., ecosystem-based flood protection, green spaces, landscape restoration, watershed protection) to create jobs, etc.
	Climate and disaster proofing of all infrastructure investment (adaptation) funded under recovery or stimulus packages
	Reorientation of public works programs to support adaptation and resilience (e.g., irrigation, reforestation)
	Investment in early warning systems (improving current, developing new systems) to build greater resilience in general (for health surveillance, but also for other types of civil emergency)
	Investment in research and development (R&D), pilot, and demonstration projects for climate- and disaster-resilient goods and services (such as new drought- and flood-resistant crops), to reduce future shocks
	Investment in building the capacity and knowledge of female farmers through women-focused investments
	Investment in sustainable urban infrastructure and services (e.g., water, sanitation, and hygiene, or WASH, and solid waste projects, to enhance health and disaster and climate resilience in urban areas, including informal settlements)
	Investment in rural support scheme spending, particularly with sustainable agriculture, to provide rural stimulus and more robust logistics and supply chains

Direct Investment, continued

	Community-targeted approaches that engage women, including adaptive safety nets (e.g., shock-responsive contingency funds and cash transfers), as part of social protection
	Development of skills development programs, including re-skilling and up-skilling of male and female workers (and especially youth) in technical and vocational education to match green and resilient investments and increase women's participation in nontraditional sectors
	Investment in rural enterprise finance and rural microfinance programs
	Investment in new information technology infrastructure to increase resilience and continued work, under crisis management and emergency response plans

Policy and Regulatory Reform	Inclusion of resilience factors (e.g., climate and disaster risk considerations) in road standards to ensure the resilience of stimulus investments
	Integration of climate and disaster risks into prudential and capital market regulations
	Introduction of financial incentive schemes (e.g., for resilient investment) to support immediate recovery and sustain benefits over the longer term
	Introduction of policy reform to enhance resilience (e.g., payment for ecosystem service schemes) and promote a sustainable recovery
	Active labor market policies and economic stimulus to strengthen job creation in resilience sectors
	Flexible and responsive labor market policies that are adaptive in times of crisis, formulated as a collaborative effort by government, the private sector, civil society organizations, and social partners

Transformational Change	Reform of government, governance, planning and policy (e.g., toward more social justice and a citizen-led transformation)
	Urban green transformation
	Transformation of rural food and land-use systems, including shift to sustainable, localized, and resilient production, and assistance in building the resilience and robustness of value and supply chains
	Rethinking of land-use development patterns, especially in highly-at-risk areas, toward resilient cities that can weather future shocks and stresses
	Well-being national accounting (e.g., away from gross domestic product)

Capacity Building/ Institutional Strengthening	Support for the developing member countries (DMCs) in developing resilience investment pipelines for the recovery packages
	Capacity building and policy support to incorporate priorities reflected in nationally determined contributions (NDCs), national adaptation plans (NAPs), and national/local disaster risk reduction plans into recovery plans
	Capacity building for mainstreaming climate and disaster resilience into national development planning (green budgeting) alongside recovery planning
	Strengthening of institutions and capabilities to assess, prepare, and reduce complex, systemic risks, including climate and disaster risks
	Capacity building for the implementation of labor market programs, integrating resilience considerations
	Improved coordination between government agencies at the national and subnational levels, and horizontally across ministries and departments, leading to setting of resilience priorities, appropriate sequencing, and policy coherence
	Increased public preparedness through better awareness of the impact of the climate crisis, disaster risks, resilience, and actions in case of an emergency
	Capacity building to increase women's resilience to climate and disaster risk

Finance	Issue of insurance-linked securities (e.g., catastrophe bonds, resilience bonds), to help raise finance for recovery plans (with resilience focus)
	Debt restructuring (e.g., debt swap examples for marine conservation and adaptation) for recovery and direct investment financing
	Use of insurance instruments (e.g., sovereign disaster insurance pools, nature-based insurance, indemnity disaster insurance, parametric disaster insurance, guarantees)
	Improved information and disclosure of climate risk (transition risk), and introduction of (mandatory) climate-related financial disclosure to enable investors to make informed decisions on post-COVID-19 investment
	Use of forecast-based finance (forecast-based triggering of the disbursement of resources to threatened communities before extreme events hit)
	Blended finance (e.g., special-purpose vehicles for climate resilience), to help raise finance for recovery packages
	Use of public financial management risk retention instruments (e.g., reserves, contingent disaster financing) to strengthen fiscal resilience in the event of shocks

B.   Low-Carbon Development

Direct Investment	Investment in low-carbon (renewable) energy production and energy storage infrastructure
	Extension and modernization of the grid to support higher renewable penetration
	Investment in electric vehicles and electric charging infrastructure, in cycling and walking infrastructure, and (post-pandemic), in low-carbon public transport
	Investment in low-carbon heat systems, including community schemes
	Investment in low-carbon research and development (R&D), pilot, and demonstration projects (e.g., battery, hydrogen, carbon capture and storage)
	Public procurement of energy-efficient appliances, lighting, and digital devices for public buildings
	Investment in large-scale landscape restoration, green space, and reforestation programs
	Development of low-carbon skills development programs, including re-skilling and up-skilling of male and female workers (and especially the youth) in technical and vocational education and training (TVET) for low-carbon and other industries, to address the immediate problem of unemployment due to the COVID-19 pandemic and structural shifts resulting from decarbonization. This intervention includes facilitating and increasing women's participation in nontraditional sectors.
	Investment in broadband connectivity and digital infrastructure to build on the global teleworking experience and encourage smart working (which indirectly contributes to low-carbon lifestyles)

Policy and Regulatory Reform	Introduction of new energy efficiency standards or regulations (e.g., for appliances, low-carbon building standards, vehicles)
	Introduction and strengthening of incentives for new climate-smart technologies (e.g., battery, hydrogen, carbon capture and storage)
	Introduction and strengthening of incentives for home renovations and retrofits, and the accelerated construction of low- and zero-energy buildings (especially public buildings)
	Introduction of credit lines, guarantees, employment support, and tax exemptions for micro, small, and medium-sized enterprises (MSMEs), including women-led MSMEs, to promote investments in low-carbon and energy-efficient measures
	Reduction or removal of fossil fuel subsidies
	Introduction of green tax regimes (e.g., carbon taxes, carbon price floor) for industry
	Introduction of financial incentive schemes (e.g., for electric vehicles) and removal of barriers (e.g., import taxes for solar energy)

Policy and Regulatory Reform, continued

	Introduction of policy reforms to support renewable energy investment (e.g., renewable energy tariffs)
	Reconsideration or reform of other subsidies to lower greenhouse gas (GHG) activities (e.g., removal of agricultural subsidies that could lead to inefficient use of resources or prevent behavioral changes that promotes resilience)
	Introduction of conditional bailouts (e.g., support for carbon-intensive industries and firms contingent on commitment to emission reduction targets or a low-carbon transition)
	Active labor market policies and economic stimulus to encourage job creation in the green sector

Transformational Change	Transformation of food and land-use systems (e.g., reducing the carbon footprint of food value chains and shifting to sustainable and low-carbon production)
	Acceleration of economic diversification (away from fossil fuels)
	Reorganization and prioritization of planning toward urban green redevelopment or regeneration and sustainable spaces (smart cities)
	Introduction of net-zero targets
	Development and scale-up of radical transport (universal and comprehensive public transport, car-free streets)
	Development of new industrial plans to promote radical restructuring

Capacity Building/ Institutional Strengthening	Support for the development of green investment pipelines for stimulus packages
	Support for the incorporation of NDCs into recovery plans (e.g., using climate investment plans linked to NDCs to identify potential investment projects)
	Strengthening of international and regional coordination and cooperation in supporting DMCs in the recovery phase
	Development of energy master plans to support a long-term shift to low- or zero-emission economies as part of COVID-19 forward planning
	Capacity building and policy support for mainstreaming climate change considerations into national development planning (green budgeting) alongside recovery planning

Finance	Increased uptake and use of green bonds, climate bonds, and sovereign blue bonds to help increase finance for recovery packages
	Additional financing instruments to support private investments in green projects such as guarantees
	Improvement of information and disclosure of climate risk (transition risk), and introduction of (mandatory) climate-related financial disclosure to enable investors to make informed decisions on post-COVID-19 investment
	Reduction of information asymmetry by improving and, standardizing metrics for the classification of assets as "green" or "brown"

References

Agrawala, S., D. Dussaux, and N. Monti. 2020. What Policies for Greening the Crisis Response and Economic Recovery? Lessons Learned from Past Green Stimulus Measures and Implications for the COVID-19 Crisis. *OECD Environment Working Papers*. No. 164. Paris: Organisation for Economic Co-operation and Development. https://doi.org/10.1787/c50f186f-en.

Allan, J., C. Donovan, P. Ekins, A. Gambhir, C. Hepburn, N. Robins, D. Reay, E. Shuckburgh, and D. Zenghelis. 2020. A Net-Zero Emissions Economic Recovery from COVID-19. *Smith School Working Paper*. 20-01. England: Smith School of Enterprise and the Environment, University of Oxford.

Asian Development Bank (ADB). 2018. *Strategy 2030: Achieving a Prosperous, Inclusive, Resilient, and Sustainable Asia and the Pacific*. Manila. https://www.adb.org/documents/strategy-2030-prosperous-inclusive-resilient-sustainable-asia-pacific.

———. 2019. *Strategy 2030 Operational Plan for Priority 3 Tackling Climate Change, Building Climate and Disaster Resilience, and Enhancing Environmental Sustainability, 2019–2024*. Manila. https://www.adb.org/sites/default/files/institutional-document/495961/strategy-2030-op3-climate-change-resilience-sustainability.pdf

———. 2020a. *ADB's Comprehensive Response to the COVID-19 Pandemic: Policy Paper*. Manila. https://www.adb.org/documents/adb-comprehensive-response-covid-19-pandemic-policy-paper.

———. 2020b. An Updated Assessment of the Potential Economic Impact of COVID-19. *ADB Briefs*. No. 133. Manila. https://www.adb.org/publications/updated-assessment-economic-impact-covid-19.

———. 2020c. *Fast-Tracking the SDGs: Driving Asia-Pacific Transformations*. Manila. https://www.adb.org/sites/default/files/publication/605796/sdgs-driving-asia-pacific-transformations.pdf.

Barbier, E. 2010. Green Stimulus Is Not Sufficient for a Global Green Recovery. *VoxEU.org*. London: Centre for Economic Policy Research. https://voxeu.org/article/urgently-needed-global-green-new-deal.

Bhattacharya, A., and J. Rydge. 2020. *Better Recovery, Better World: Resetting Climate Action in the Aftermath of the COVID-19 Pandemic*. Report prepared for the Coalition of Finance Ministers for Climate Action. July. Washington, DC.

Büchs, M., M. Baltruszewicz, K. Bohnenberger, J. Busch, J. Dyke, P. Elf, A. Fanning, M. Fritz, A. Garvey, L. Hardt, E. Hofferberth, D. Ivanova, A. Janoo, D. O'Neill, M. Guillen-Royo, M. Sahakian, J. Steinberger, K. Trebeck, and C. Corlet Walker. 2020. Wellbeing Economics for the COVID-19 Recovery: Ten Principles to Build Back Better. *WEAll Briefing Papers: Short Summaries of Big Issues*. Glasgow: Wellbeing Economy Alliance.

Carnell, R., Sakpal, P., Pang, I., Mapa, N., and Patterson, W. 2020. *Asia's lamentable green response to Covid-19* ING. Singapore. https://think.ing.com/uploads/reports/Asias_green_response_100820_AOT.pdf

Climate Action Tracker (CAT). 2020. *A Government Roadmap for Addressing the Climate and Post COVID-19 Economic Crises* (April 2020 update). https://climateactiontracker.org/publications/addressing-the-climate-and-post-covid-19-economic-crises/.

Climate Council and AlphaBeta Australia. 2020. *Clean Jobs Plan.* https://www.climatecouncil.org.au/wp-content/uploads/2020/07/Climate-Council_AlphaBeta-Clean-Jobs-Plan-200720.pdf.

Coady, D., I. Parry, L. Sears, and B. Shang. 2015. How Large Are Global Energy Subsidies? *IMF Working Paper.* Washington, DC: International Monetary Fund. https://www.imf.org/en/News/Articles/2015/09/28/04/53/sonew070215a.

Dervis, K., and S. Strauss. 2020. The Carbon Tax Opportunity. 6 May. Washington, DC: The Brookings Institution. https://www.brookings.edu/opinions/the-carbon-tax-opportunity/.

Energy Transitions Commission (ETC). 2020. Help the Global Economy Recover while Building a Healthier, More Resilient, Net-Zero Economy. London. http://www.energy-transitions.org/sites/default/files/COVID-Recovery-CoverLetter.pdf.

European Bank for Reconstruction and Development (EBRD). 2020. The EBRD and the Coronavirus Pandemic. https://www.ebrd.com/what-we-do/coronavirus.

G20. 2017. G20 Leaders' Declaration: Shaping an interconnected world. Hamburg. http://www.g20.utoronto.ca/2017/2017-G20- leaders-declaration.pdf

Garrett-Peltier, H. 2017. Green versus Brown: Comparing the Employment Impacts of Energy Efficiency, Renewable Energy, and Fossil Fuels using an Input-Output Model. *Economic Modelling.* 61(C). pp. 439–447. Elsevier.

Gentilini, U., M. B. A. Almenfi, P. Dale, A. V. Lopez, I. V. Mujica Canas, R. E. Q. Cordero, and U. Zafar. 2020. Social Protection and Jobs Responses to COVID-19: A Real-Time Review of Country Measures (8 May 2020). *COVID-19 Living Paper.* Washington, DC: World Bank Group.

Gentilini, U., M. B. A. Almenfi, I. Orton, and P. Dale. 2020. *Social Protection and Jobs Responses to COVID-19: A Real-Time Review of Country Measures.* Version 6 (24 April). Washington, DC: World Bank.

Global Carbon Project (GCP). 2011. Global Emissions Rebound to Record Levels after GFC. 5 December. Canberra. https://www.globalcarbonproject.org/carbonbudget/archive/2011/CICERO_Budget2010_final.pdf.

Global Commission on Adaptation (GCA). 2019. *Adapt Now: A Global Call for Leadership on Climate Resilience.* Washington, DC: World Resources Institute.

Global Commission on the Economy and Climate (GCEC). 2014. *Better Growth, Better Climate: The New Climate Economy Report.* Synthesis Report. Washington, DC. https://sustainabledevelopment.un.org/content/documents/1595TheNewClimateEconomyReport.pdf.

———. 2020. *NCE Key Messages Pack: Special Edition on COVID-19.* Washington, DC.

Gosens, J., and F. Jotzo. 2020. How Green is China's Post-COVID-19 'New Infrastructure' Stimulus Spending? *East Asia Forum.* 5 May. Canberra: Crawford School of Public Policy, Australian National University. https://www.eastasiaforum.org/2020/05/05/how-green-is-chinas-post-covid-19-new-infrastructure-stimulus-spending/.

Green Stimulus Proposal. 2020. A Green Stimulus to Rebuild Our Economy: An Open Letter and Call to Action to Members of Congress. https://medium.com/@green_stimulus_now/a-green-stimulus-to-rebuild-our-economy1e7030a1d9ee%0D (accessed 27 April 2020).

Hallegatte, S., J. Rentschler, and J. Rozenberg. 2019. *Lifelines: The Resilient Infrastructure Opportunity.* Washington, DC: World Bank.

Hammer, S., and S. Hallegatte. 2020. Planning for the Economic Recovery from COVID-19: A Sustainability Checklist for Policymakers. *World Bank Blogs.* 14 April. https://blogs.worldbank.org/climatechange/ thinking-ahead-sustainable-recovery-covid-19-coronavirus.

Hepburn, C., B. O'Callaghan, N. Stern, J. Stiglitz, and D. Zenghelis. 2020. Will COVID-19 Fiscal Recovery Packages Accelerate or Retard Progress on Climate Change? *Smith School Working Paper.* 20-02. England: Smith School of Enterprise and the Environment, University of Oxford.

International Energy Agency (IEA). 2020a. Global Energy and CO_2 Emissions in 2020. *Global Energy Review 2020.* April. Paris. https://www.iea.org/reports/global-energy-review-2020/global-energy-and-co2-emissions-in-2020.

———. 2020b. Green Stimulus after the 2008 Crisis: Learning from Successes and Failures. 29 June. Paris. https://www.iea.org/articles/green-stimulus-after-the-2008-crisis.

———. 2020c. *Renewable Energy Market Update (Outlook for 2020 and 2021): COVID-19 Impact on Renewable Energy Growth.* Paris. https://www.iea.org/reports/renewable-energy-market-update.

International Federation of Red Cross and Red Crescent Societies (IFRC). 2010. Mangrove Plantation in Viet Nam: Measuring Impact and Cost Benefit. *Case Study.* Geneva. https://www.ifrc.org/Global/Publications/disasters/reducing_risks/Case-study-Vietnam.pdf.

International Monetary Fund (IMF). 2020. Greening the Recovery. *Special Series on Fiscal Policies to Respond to COVID-19.* 20 April. Washington, DC: IMF Fiscal Affairs Department.

———. 2018. *Public Investment Management Assessment – Review and Update.* Washington, DC. https://www.imf.org/en/Publications/Policy-Papers/Issues/2018/05/10/pp042518public-investment-management-assessment-review-and-update

International Renewable Energy Agency (IRENA). 2020. How Falling Costs Make Renewables a Cost-effective Investment. 2 June. Abu Dhabi. https://www.irena.org/newsroom/articles/2020/Jun/How-Falling-Costs-Make-Renewables-a-Cost-effective-Investment.

The Investor Agenda. 2020. A Sustainable Recovery from the COVID-19 Pandemic. https://theinvestoragenda.org/wp-content/uploads/2020/05/FINAL-THE-INVESTOR-AGENDA_-A-SUSTAINABLE-RECOVERY-FROM-COVID-19-CLEAN_v1.pdf.

Jacobs, M. 2012. Green Growth: Economic Theory and Political Discourse. *Centre for Climate Change Economics and Policy Working Paper No. 108, Grantham Research Institute on Climate Change and the Environment Working Paper No. 92.* London. Grantham Research Institute on Climate Change and the Environment, London School of Economics and Political Science.

Jotzo, F., T. Longden, and Z. Anjum. 2020. Fiscal Stimulus for Low-Carbon Compatible COVID-19 Recovery: Criteria for Infrastructure Investment. *CCEP Working Paper 2005.* June 2020. Canberra: Centre for Climate and Energy Policy, Crawford School of Public Policy, Australian National University.

Levy, J., C. Brandon, and R. Studart. 2020. Designing the COVID-19 Recovery for a Safer and More Resilient World. *WRI Commentary.* 14 May. Washington, DC: World Resources Institute.

Mechler, R. 2016. Reviewing Estimates of the Economic Efficiency of Disaster Risk Management: Opportunities and Limitations of Using Risk-Based Cost–Benefit Analysis. *Natural Hazards.* 81. pp. 2121–2147. https://doi.org/10.1007/s11069-016-2170-y.

Mission 2020. 2020. Green Stimulus: Case Studies from 2008–2009. April. https://mission2020.global/wp-content/uploads/Green-stimulus_-case-studies-from-2008-2009.pdf.

Mountford, H. 2020. *Responding to Coronavirus: Low-carbon Investments Can Help Economies Recover.* Washington, DC: World Resources Institute.

Network of Central Banks and Supervisors for Greening the Financial System (NGFS). n.d. Accessed 10 August 2020. https://www.ngfs.net/en.

Organisation for Economic Co-operation and Development (OECD). 2020. *Building Back Better: A Sustainable, Resilient Recovery after COVID-19.* Paris.

Oxfam International. 2020. Dignity Not Destitution: An 'Economic Rescue Plan for All' to Tackle the Coronavirus Crisis and Rebuild a More Equal World. *Policy Paper.* 9 April. Nairobi.

Pinner, D., M. Rogers, and H. Samandari. 2020. Addressing Climate Change in a Post-pandemic World. *McKinsey Quarterly.* 7 April. New York: McKinsey & Company.

Regalado, F. 2020. Asia Risks Missing 'Green' Economic Reset after Coronavirus. *Nikkei Asian Review.* 23 June. https://asia.nikkei.com/Spotlight/Asia-Insight/Asia-risks-missing-green-economic-reset-after-coronavirus.

Sayeh, A., and R. Chami. 2020. Lifelines in Danger. IMF.org. https://www.imf.org/external/pubs/ft/fandd/2020/06/COVID19-pandemic-impact-on-remittance-flows-sayeh.htm?utm_medium=email&utm_source=gov%E2%80%A6.

United Nations Department of Economic and Social Affairs (UN/DESA). 2020. The COVID-19 Pandemic: A Speedy and Balanced Recovery of Europe Will Remain Critical for the World to Return to the Trajectory of Sustainable Development. *Policy Brief.* no. 63 (1 May). New York.

United Nations Development Programme (UNDP). 2020a. Climate 2020. http://www.undp.org/content/undp/en/home/events/2019/climate-2020.html.

———. 2020b. The Social and Economic Impact of COVID-19 in the Asia-Pacific Region. Position Note. 28 April. Bangkok: UNDP Regional Bureau for Asia and the Pacific.

United Nations Environment Programme (UNEP). 2019. UNEP Emissions Gap Report 2019. Nairobi. https://www.unenvironment.org/resources/emissions-gap-report-2019.

Vivid Economics. 2020. Case Study: Greenness of Stimulus Index. https://www.vivideconomics.com/casestudy/greenness-for-stimulus-index/.

www.ingramcontent.com/pod-product-compliance
Lightning Source LLC
Chambersburg PA
CBHW052043190326
41519CB00003BA/260